YOUR KNOWLEDGE HAS VALUE

- We will publish your bachelor's and
  master's thesis, essays and papers

- Your own eBook and book -
  sold worldwide in all relevant shops

- Earn money with each sale

Upload your text at www.GRIN.com
and publish for free

Bibliographic information published by the German National Library:

The German National Library lists this publication in the National Bibliography; detailed bibliographic data are available on the Internet at http://dnb.dnb.de .

This book is copyright material and must not be copied, reproduced, transferred, distributed, leased, licensed or publicly performed or used in any way except as specifically permitted in writing by the publishers, as allowed under the terms and conditions under which it was purchased or as strictly permitted by applicable copyright law. Any unauthorized distribution or use of this text may be a direct infringement of the author s and publisher s rights and those responsible may be liable in law accordingly.

Imprint:

Copyright © 2020 GRIN Verlag
Print and binding: Books on Demand GmbH, Norderstedt Germany
ISBN: 9783346113894

This book at GRIN:

https://www.grin.com/document/516556

Nithish Reddy, Ms. Keerthi

Recent technological advancements in the refrigeration technology

GRIN Verlag

GRIN - Your knowledge has value

Since its foundation in 1998, GRIN has specialized in publishing academic texts by students, college teachers and other academics as e-book and printed book. The website www.grin.com is an ideal platform for presenting term papers, final papers, scientific essays, dissertations and specialist books.

Visit us on the internet:

http://www.grin.com/

http://www.facebook.com/grincom

http://www.twitter.com/grin_com

RECENT TECHNOLOGICAL ADVANCEMENTS IN REFRIGERATION TECHNOLOGY

A Quick Review

P. NITHISH REDDY

Table of contents

CHAPTER-1

1.1 Effective expansion technology

improves expansion power and energy efficiency of vapour compression refrigeration systems. Murthy et al.[1] conducted an extensive review on the use of variety of expander mechanisms such as reciprocating piston, rolling piston, rotary vane, scroll, screw and turbine and their performance is reported. They observed that the trans-critical CO_2 has reported COP of 30% whereas non - CO_2 systems have registered their COP as 10%.The maximum expander efficiency was 83% obtained by the use of scroll expander in the CO_2 refrigeration cycle. Further it is understood that recuperation of the energy of the expanders is an encouraging practice of refrigeration systems. Research on the topics such as irreversibility, heat transfer ,expansion process, internal leakage etc. is still open.

1.2 Pinch technology

plays important role in liquefaction cycle for cryogenic processes. Lei et al.[2] showed that the cascade expanders can influence the temperature characteristics of hot and cold streams for transfer of heat in cryogenic systems. Depending on the pinch technology *a novel intercooled series of expanders* has been developed. Pinch technology make provisions to analyse different ways of plans for large scale cryogenic systems. Heat exchanger and expander in this expansion refrigeration system are joined in series. Heat transfer efficiency found higher for ISE cryogenic systems , use of pinch technology when compared to common traditional methods could save energy of about 30% to 50%.

1.3 Bubble absorption refrigeration technology

was studied and the methods for improving the performance was evaluated by Wu et al.[3]. In this, bubble refrigeration systems and bubble absorbers were used. Bubble behaviour properties were observed and explained graphically. The present refrigerant absorption pairs in bubble absorption refrigeration were explained, in this there were fifty fluids included. This study is helpful for the research people in the fields of absorption refrigeration technology. Performance enhancement can be achieved in three different types of methods such as mechanical-physical method, chemical method and adscititious contactless forces method. Ultrasonic field and magnetic field are explored in adscititious contactless forces method. Impact mechanism and dependability is discussed in order to show the effect of improvement. Current and potential fluids have been examined carefully.

1.4 Active off-grid refrigeration technology

for food preservation is the most neglected in India due to insufficient access to the different forms of energy. Aste et al. [4] carried out great study on the review of literature and scientific technologies in this case. The main aim is to achieve preservation of different kinds of foods and the next to be concentrated on technical, economic and social point of view. In areas without power grid food preservation is done through active/passive refrigeration that is achieved with the use of electricity. In commercial cases off- grid active refrigeration technology is stimulated. Innovative refrigeration systems have been implemented for describing relevant experiences.

1.5 Biomethane liquefaction technology

is an alternative refrigeration technology which is gaining importance in recent days. Capra et al. [5] showed the comparison of economical and technical analysis of alternative refrigeration technology for the development of liquified biomethane, biofuel which is also termed as liquified bio LNG. These have been developed in order to design biogas plants in which they are installed at the downstream of biogas upgrading step so that the biomethane is obtained in liquid state at a temperature of -152 ^{0}C AND 2 bar pressure. Five types of technologies have been considered in this, starting with liquid nitrogen vaporization (benchmark); reverse Rankine cycle with mixed refrigerant; reverse Brayton cycle; Claude cycle; reverse Stirling cycle. Use of mixed refrigerants has been the better choice taken in the Rankine cycle. The primary energy consumption has been confined between14% and 20%.

1.6 TES-backed up vapour compression refrigeration system

was modeled and analyzed *by* Rodriguez et al. [6]. *They explained the* power control and optimization of a thermal energy storage(TES) system in connection with a vapour compression refrigeration facility depending upon phase change materials(PCM).A unique plan of PCM dependent TES tank and its linkage with the present refrigeration system is designed. In this the combined dynamic model is examined by taking a look at the various time frames that coincide at the interconnected system. Different kinds of operating techniques have been explained depending upon deliberate use of TES tank as a low temperature energy buffer to separate cooling urge and production, considering that the stable feature and power limits. The cooling power loops were analyzed in detail in this work and the power monitoring control was solved.

1.7 Nanotechnology for refrigeration

was introduced keeping in mind the energy usage and environmental concerns, this technology enhances the thermal performance of the refrigeration system. This can be done either by changing the properties of the working fluids or by using different systems. Now a days nanofluids have got much focus due to their outstanding thermophysical properties that can be pre owned comfortably in refrigeration and air conditioning systems. Many future studies are requirement for further enhancement of their properties. Low boiling point nanofluids need to be synthesized. These days industries are showing much interest towards natural refrigerants keeping in mind about the pollution they cause. Nanofluids surface tension and dielectric properties needs further study. Different kinds of water dependent brines are used as secondary refrigerant but there is no study conducted on this water based nanofluids[7].

CHAPTER -2

2.1 A Combined power and refrigeration cycle

depends upon ocean thermal energy conversion system eg; Kalina cycle. In this kind of cycle the solar energy is used to increase the heat source temperature. Chilled volume is parallelly obtained by the ejector refrigeration cycle that mainly concentrates on the exhaust heat that can be collected from turbine in Kalina cycle. In Kalina cycle and ERC the working fluids that are selected are Ammonia-water and isobutane. In order to calculate the performance ,a newly proposed cycle is kept in connection with the commercial ASPEN PLUS. The primary energy ratio is more for the proposed cycle when compared with the Kalina cycle i.e.,29.2% since having lesser power output.T he exergy efficiency and the primary energy has a highest value of about 41.88% and 8.3%[8].

2.2 Organic absorption refrigeration processes

was thoroughly reviewed Papadopoulos et al. [9].Here use of natural organic mixtures as working fluids in absorption system for generation of cooling are vital since they allow development of fresher and renewable energy sources. The system can be improved by arranging the past works that have been done in organic fluids in a proper way that have more importance than the conventional inorganic fluids. The new enhancements have divided the working fluids in single effect cycles and double effect. The proper reports have been submitted relating to the operating conditions tested, the basis on which the working fluid has been selected. Thermodynamics of the working fluids that suits the ABR process are also been tested. The major practical works are done relating to the single effect systems that mainly exposes to the advancing compound process alterations. Thermodynamic interrogations majorly unite experiments with parameter approximation for model enhancement[9].

2.3 Biodegradable Refrigeration lubricant

are been researched for enhancing the performance of the compression system which in turn improves the Coefficient of performance. In case of refrigeration oil the important thing to be considered is the affinity of piston ring tribo pair of the hermetically covered compressor[10]. The base oil that used in the preparation of refrigeration oil has obtained from petroleum products. The scarcity of petroleum reserves and increased prices prices of crude oil has made to look for eco friendly energy resources to synthesize refrigeration oil. The existing exploration intends to formulate a rapeseed oil based trimethylolpropane triester as biodegradable refrigeration oil with efficient thermo oxidative stability and cold flow nature.

Differential Thermal Analyses (DTA) and Thermo gravimetric Analyses are done to examine the work of thermo-oxidative stability of refrigeration oil[10].

2.4 Transcritical R744 refrigeration systems are the remarkable technological improvements observed in the market of modern European refrigeration systems. The working of state -of -the art pure refrigeration plants has been vastly examined for a medium range food retail situated in European climate conditions. The outcomes disclosed that when collated to R404A expansion unit ,the energy efficiency accomplished with the R744 refrigerating systems has increase in exit temperature. Additionally in view of the solutions obtained the progressed commercial CO_2 gives energy savings from 3% to 37.1% all over Europe when compared to R404A.Finally it has been clearly understood that transcritical R744 ejector assisted parallel systems have been the highest well organized and ecological friendly technologies for European retail food industry[11].

2.5 Direct expansion geothermal refrigeration system

using CO_2 transcritical cycle is one of the energy efficient refrigeration systems. In this type of geothermal refrigeration system ,the refrigerant is being kept under the ground so that it exchanges heat with the soil. Gao, et al. [12] studied the performance of the system through mathematical models. Static Distribution Parameter Method and the Lumped Parameter Method are used in order to describe different processes of thermodynamics. The   simulation of this geothermal refrigeration is done by using these mathematical models . By using these simulation models. the maximum COP recorded is 0.966 and the optimum side pressure recorded is 14.6MPa.

2.6 Hybrid dual-temperature absorption refrigeration system

was investigated by Mohammadi et al. [13]. The implementations of multi evaporator and lower temperature refrigeration systems are growing very fast with main focus on new features that are economic and environment friendly. In this two compressors are employed between evaporators and absorbers in order to increase the absorber pressure to the maximum. A hybrid dual evaporator and absorber system is used for refrigerating and freezing implementations are calculated from thermodynamic, environmental and economic point of view. The comprehensive work shows that by employing a compressor between the high temperature evaporator and absorber is used in increasing the COP very dominantly. Economic studies shows that the compressor should be fixed properly since the design and cost of the operation may overcome the COP of the system.

2.7 Ejector sub cooled refrigeration system

is initiated with a recently developed cycle of configurations in order to minimise the throttling loss and upgrade compressor pressure. The major benefit of this is that it does not need any vapour-liquid filter. Vapour compression refrigeration cycle and standard expansion are differentiated using exergy and energy techniques and different variables have also studied carefully. The outcomes indicate that the COP, volumetric efficiency, Exergic efficiency increases. The COP at a particular conditions that is obtained as 17.4% and 26.6%.All these parameters have been calculated at low temperatures only. Exergic efficiency has been improved from 17.9% to 22.2%.

CHAPTER -3

3.1 Cascade absorption refrigeration (CAR)

process is used to recover low-grade waste heat. In China, about 50% of the industry's energy is discarded, the largest of which is waste heat .A large amount of waste heat is absorbed through the skin in the industrial process before being used effectively. The macro figures show that the annual waste heat loss in northern China is around 2.6 billion tons.The computation procedure by Cui et al.[15] revealed that average annual expenditures is 11% far more than the minimum result, that implies that the thermal efficiency of the scheme is small. The energy, exergy and economic (3E) analysis of the cascade refrigeration system were conducted depending on the computational outcomes[15].

3.2 Ejector expansion refrigeration

system with booster compressor was thermodynamically analyzed and optimized by Rostamnejad et al.[16]. This research recommends, examines and improves a novel ejector-expansion refrigeration cycle (EERC). The current system is modified taking into consideration the exergy efficiency as the main factor as well as at optimal circumstances a correlation is created with that of the conventional EERC and typical vapor compression refrigeration system. Using a particular procedure of constant pressure blending for simulating the ejector a comprehensive and detailed thermodynamic study is carried out on the basis of the first and second laws and the capacity of the device for six necessary refrigerants is investigated. The findings indicate that of all the six refrigerants examined, R1234ze is the best for which new system has exergy efficiency values of 5.7% and 15.5% greater than the standard EERC and conventional vapor compression system.

3.3 Heat driven absorption refrigeration technology

includes refrigeration absorption technology that is thermally operated[17]. The system evaluation of the absorption refrigerator was achieved using the heat-driven absorption mechanism to recycle the excess heat and quantitative analysis from the industry. Findings shows that when the evaporating temperature is 5 ° C, the maximum COP of the absorption sump is nearly 0.825 below 60 degrees C generator temperature. Using the thermally-driven absorption method to recycle the industry's excess heat and quantitative analysis, device evaluation of the absorption refrigerator was achieved. Annual energy savings fall slightly from £ 21 to £ 19.8 with a rise in the temperature of the installed generator from 53 ° C to 90 ° C.

3.4 Hybrid solar-assisted power refrigeration system

working with supercritical carbon dioxide was optimized with a Tri-objective optimization method by Khanmohammadi et al.[18]. The objective of this research is to maximize a current power-refrigeration system assisted by a parabolic trough solar collector. The yields of tri-objective computation show that the total amount of exergy destruction declined by 436.7 kW with regard to the basic system. Solar energy provides the system's required thermal energy requirement using a PTSC. For the optimization process, an economic analysis of the proposed method is carried out. The groundbreaking system includes CO_2 as a working fluid, a natural medium comprising of three sub-cycles: a cycle of Brayton, a cycle of Rankine, and a cycle of refrigeration of vapor compression. The key benefits by using CO_2 as working fluid are; energy-friendly properties such as limited global warming potential and zero-ozone depletion are present in large amounts, and excellent thermodynamic properties.

3.5 Hybrid refrigeration system for industrial waste recovery

The integrated system consists of an Organic Rankine cycle, a vapour compression cycle and a liquid desiccant unit. The unit can realistically convert 200 kW of heat source to 50 kW of selective cooling and 132 kW of latent cooling effect. The critical cooling can be provided by the cooling and VCC unit. Four working fluids were selected to test the efficiency under specific evaporating temperatures of the ORC based hybrid refrigeration system. The method using n-butane as the ORC working fluid would be able to produce maximum amount of latent cooling effect under the same working conditions[19].

3.6 Diffusion-absorption refrigeration technologies

use variety of ammonia with hydrogen as a supplementary gas[20]. The main factors that affect energetic efficiency are flux regime, activation thermal diameter. The outcome is confined as technology is interesting because it has more profits when that is compared with the vapour compression of low cooling volume since they would not contain moving parts. Various kinds of mixtures were examined in order to increase the efficiency. The major drawback of this is the research has been carried on advanced configurations in order to improve efficiency. The energy savings can be done is upto 20%.

3.7 Heat-driven ejector refrigeration systems

became an alternate option for the conventional compressor-based cooling technologies for reducing the demand in the consumption of energy. A detailed study has been done on working fluids and refrigeration systems[21]. An in-depth analysis has been done for ejector technology and performance of properties of the refrigerants. The Study is carried out in four fragments. In the first it has been discussed about the ejector technology and in the second one about the properties of refrigerants and their effect on ejector performance. The third part concentrates on the main jet refrigeration cycle. All ejector technologies have been used maintaining the relation between the past, present and future. Additional work is carried out on-design and off design working situations using experimental and mathematical calculations and investigations.

3.8 Fishing vessel refrigeration systems

are driven by exhaust heat from engines[22]. Though, there can be obstacles in boosting the COP / EER of these heat-driven cooling systems and preserving their functional sustainability under intense ocean conditions. A hybrid heat-driven cooling system is then adopted, incorporating the benefits of various types of projects. Furthermore the fishing vessel would help us achieve fuel savings. For a fishing vessel, the energy efficiency of its diesel engine is only 35–40 %, with more than fifty percent of the energy being consumed as exhaust energy, fishing vessel cooling systems operated by exhaust heat from engine exhaust heat drained by jacket water, cooling air and exhaust gas. Due to its low energy efficiency for fishing vessel engines, the recovery of excess heat through engines provided an efficient method that can save energy in fishing vessels, making the need for cooling systems driven by exhaust heat from ice-making engines, cooling and air conditioning on fishing vessels a unique project.

3.9 Phase change Materials for Food Refrigeration Applications

Corn oil ester has never shown damaging tendencies but is also non-toxic[23].In regards, corn oil ester and water solutions deliver a constant distribution as well as cheaper option especially in comparison to fatty acid esters commonly used for the programs below 0 ° C. Corn oil ester in tap water blends was studied for the advancement of phase transition materials as heat energy storage media which could be used for food temperature cooling systems. The analysis also noticed that there can be a low or even no or insignificant degree of cooling for the PCM applicants during experimental conditions. Such properties enable PCMs with large heat energy and appropriate phase change temperatures for medium- and low-temperature food refrigeration purposes.

A gas compression refrigeration cycle for fresh water production from atmospheric air was thermodynamic analyzed by Zolfagharkhani et al.[24] Many research activities have been conducted in recent decades to find creative solutions to deliver new groundwater supplies or increase the use of current water resources. The water vapor content of air is another one of those new assets. The aim of Zolfagharkhani et al.[24] study was to analyze the production of liquid water from atmospheric water vapor with the aid of a cooling process. The results suggest that a household device size would produce 22–26 l / day of fresh water while the energy intensity differs between 220 and 300 Wh / l. The quantity of water supply and energy strength is evaluated under different environmental conditions.

CHAPTER 4

4.1 Absorption chillers

is a heat-driven absorption freezing climate friendly technology. Here article [25]provides outcomes of a socio-technical configuration analysis that describes hurdles and engines of secured delivery in a more systematic way of creative small-temperature absorption chillers.

Mylona et al. [26] carried out a comparison work on four different configurations of the industrial refrigeration system. The prototype is verified for both energy and environmental situations against real monitoring data. The baseline design is used in a real case study convenience store that has a hold-alone system of reference refrigeration. The weather assessment was conducted for the London DSY file.

Prepare to identify the possibility of colder than unusual year results in controlled service of refrigeration systems for a cooler than normal year. This program ended with a 17.4% drop in the case study store's total annual energy consumption. Since the CO_2 booster device output is decreased as the outdoor temperature increases, the weather conditions in London are not stringent as the outdoor temperature does not exceed 27 ° C for most of the time throughout the year.

4.2 Solar sorption refrigeration technologies:

The technology history and recent advances in solar sorption cooling technologies are recorded after an application of basic principles. Solar-powered absorption cooling systems are appealing solutions which would not only meet the needs for air conditioning, refrigeration, ice production purposes, but can also fulfill demand for energy saving and protection of the environment. However due to the relatively poor COP or low energy conversion performance of solar collectors, most implementations are still being tested and developed and tested[27]. One limitation to practical adsorption refrigeration systems is the degradation by adsorption of adsorbents. Destitute issue with heat and mass transport is a barrier to avoid changes in the strategy of adsorption cooling. There are two methods of maximizing heat and mass transfer. Adsorption cooling would be a potential alternative to absorption cooling for future refrigerant substitution. And it is necessary to try to bring further adsorption refrigeration systems into operation[28].

4.3 Solar-powered closed sorption refrigeration systems

operating pairs (fluids) for solar-closed sorption (absorption and adsorption).Typically, adsorption cooling requires a low temperature source of heat than the absorption cooling requirement. The history of development and recent advances in solar sorption after the basic principles of these structures have been applied. Heat driven cooling technologies are divided into two categories: heat transfer-mechanical technology and sorption technology (open systems or closed systems). The adverse environmental impact of burning fossil fuels has prompted the energy research community to take serious consideration of renewable sources, such as natural solar power. Solar-powered closed sorption cooling technologies can be attractive alternatives not only to meet the requirements for air conditioning, cooling, ice making, thermal energy storage or hybrid heating[29].

4.4 Solar-driven ejector refrigeration technologies:

The technology history and recent developments in solar-driven ejector cooling systems are recorded and classified by Abdulateef et al. [30]. A lot of research work to be done for large-scale industrial applications and to replace modern cooling devices. Solar-driven ejector cooling technologies not only are capable of serving cooling requirements such as air conditioning and ice-making, and preserving medical or food in remote areas, yet energy saving and environmental protection standards can also be fulfilled. Different types of operating fluids can be used in the cycle; each provides different efficiency and functionality. The selection of a working fluid involves many problems, especially physical and thermodynamic characteristics.

Sorption devices such as absorption chillers, adsorption freezers or mild detergent wheels are the most common solar cooling systems. Solar cooling is a strategy that helps to reduce the grid's electricity demands throughout the summer period. In addition, the high capacity of solar energy in a wide variety of generals makes it a vital source of renewable energy for the current state of the world. Solar power use is one of the most effective ways to address multiple threads such as climate change, rising demand for energy, and high electricity prices. The device is configured in each case by choosing the optimum configuration of the ejector and the optimum temperature of the generator[31].

4.5 Magnetocaloric refrigeration effect:

The magneto caloric effect and its simplest implementation, magnetic refrigeration, are issues of specific interest because of the improvement of cooling and temperature control systems . Energy efficiency, in conjunction with other environmental impact correlated with a technology that would not rely on harmful gas compression / expansion are the huge benefits. The research in this area focused on basics of the impact, the methods for its calculation, taking into account potential objects found in sample identification, a detailed and comparative study of various magnetocaloric materials, as well as possible routes for enhancing their performance[32]. On the research side of products, one of the main lines is the creation of new processes with major magnetocaloric reactions. Magnetic refrigeration is an interesting research field with specific interest both in the search for new magnetocaloric materials .

4.6 Cascade Vapour Refrigeration Systems

shows a way for improvement in the efficiency and flow rate with an increase in compressor activity and evaporator temperature. Refrigerants such as R134a / R23, R410A / R23 and R404A / R170 were successfully tested at a combination between 10 ° C and 5 ° C overheating and sub-cooling[33].The condenser temperature difference in the high temperature zone varied from 30 to 50 ° C, while the evaporator temperature ranged from -70 ° C to -50 ° C in the low temperature region..The R134a / R170 refrigerant pair was found to have the highest COP and the lowest mass flow rate among all the refrigerant pairs tested[33].

CHAPTER-5

5.1 An Integrated Kalina and absorption refrigeration

cycles for simultaneously cooling and power generation was done by Dhahad et al. [34]. In addition, taking into account mathematical modeling, the parametric examination is carried out to evaluate the effect of the key parameters on the product's total unit cost. Minimal-temperature heat sources such as geothermal sources may be a feasible option for producing the required power, cooling, cooking, and other by-products. The combined cycle's energy and exergy efficiencies and net capacity is 41.33%, 27.47% and 158.3 kW respectively. The average unit value of the cooling, electricity and overall network were calculated at $148.5/GJ, $97.16/GJ, and $19.44/GJ, accordingly.

5.2 A Multi-effect desalination unit equipped with a cryogenic refrigeration

system was studied by Ghorbani et al.[35]. A natural-gas-fired power plant is the first option, whereas the second is focused on a steam-solar power plant using dish collectors. Two options were considered and contrasted to provide the device with power generation and the necessary heat to operate the absorption process. The extension of sensitivity analysis on the essential parameters of the developed system indicates an effective technique for modeling improvement of device parameters. The power generated per kg of LNG is approximately 0.19 kWh, that is less than or equal to the other relevant patents. In the heat exchangers, the most thermodynamic irreversibilities occur in the heat exchangers (61 percent), although they have the lowest efficiency of exergy compared to other plant equipment.

Ghoreishi et al.[36] compared the pros and corns of Ice versus battery storage or refrigeration systems for remote sites. National tendency towards renewable energy projects promotes the implementation of different scales of alternative energy systems. The technology of ice storage can be helpful in raising the maximum capacity of the cooling system. It is demonstrated that, due to lower initial investment, ice storage technology can be financially more feasible than battery storage technology for mine cooling purposes . In relation, ice storage has enhanced the investment's payback period by allowing decision-makers to reduce their plant capacity and save from the associated capital investment.

5.3 A Two Stage Compression-Absorption Refrigeration System

for Ice Cream Hardening Plant was presented by Patel et al.[37]. They explored a new theory of the Two Step Vapor Compression-Absorption Cascade Refrigeration System (TSVCACRS) for obtaining industrial cooling at low temperatures. The findings suggest that the planned TSVCACRS system would reduce compressor operation by up to 28% compared to an actual TSVCRS mounted. In fact, from an energetic and exergetic point of view, the optimal generator temperature for the current system is considered to be 85 ° C. The heat exchanger of the cascade condenser operates as an evaporator for VARS and the TSVCRS condenser. The system consists of a Two Stage Vapor Compression Refrigeration System (TSVCRS) with a single stage Vapor Absorption Refrigeration System (VARS) integrated flash intercooler; thermally connected by a cascade condenser heat exchanger.

5.4 Simulation Toolbox (DSST)

is used to optimize the lifecycle of industrial refrigeration systems[38]. A creative simulation toolbox has been introduced, enabling designers to be better supported during the development process. This would provide a real-time analysis of magnetic cooling system-related ecological factors. The Simulation Toolbox was evaluated in a specific industrial environment after the deployment process .This helps each participant to have a dedicated and consumer-friendly modeling toolbox that assists developers in designing creative products with a real-time analysis of environmental and economic indexes and diagrams. For each of the main components that make up a magnetic refrigerator, distinct Toolboxes were developed. Moreover, the logic behind the LCA and LCC standards is the same. Throughout system level appraisal, all official data relating to different components from the different measurement instruments will be compiled into a standard (web-based)[38].

5.5 An alternative food transport refrigeration systems

is first tested at component level of a revolutionary simulation toolbox[39]. Simulation Toolbox provides manufacturers with a real-time analysis of economic and environmental indices and graphs throughout the production of creative components. As can be seen from the study, for each of the key components that make up a magnetic refrigerator, separate Tool boxes were made. However, the reasoning behind the LCA and LCC values is the same. While system level appraisal, all economic data relating to different components from the different measurement instruments will be compiled into a standard (web-based) workspace, allowing for better use from the perspective of the household suppliers. The working costs are quite high in all the three methods.

5.6 Solar-driven hybrid absorption-thermochemical refrigeration system

was studied by Fito et al. [40] . The performance-storage relationships are evaluated for the $NH_3/BaCl_2$ pair in terms of variations in operating conditions and several reactive composite implementation parameters. The performance of the hybrid system with the ($NH_3/LiNO_3$ + $NH_3/BaCl_2$) pair combination is compared to its subsystems against a variable demand profile calculated from climatic data of July in Barcelona, Spain. The concurrent parametric study suggests that the $NH_3/LiNO_3$ absorption subsystem approaches close to the average COP under the conditions stated, and the thermochemical subsystem provides its highest COP with the $NH_3/BaCl_2$ pair. Hybridization is a technology with a lot of ability, very young. This makes it possible that emerging technologies to incorporate the advantages while canceling or at least reducing the drawbacks, and the number of possible combinations is significant.

CHAPTER-6

References

1. Murthy, A. A., Subiantoro, A., Norris, S., & Fukuta, M. (2019). *A Review on Expanders and their Performance in Vapour Compression Refrigeration Systems. International Journal of Refrigeration, Vol. 106, 2019, pp 427-446. https://doi.org/10.1016/j.ijrefrig.2019.06.019.*

2. *Lei, G., Yu, J., Kang, H., Li, Q., Zheng, H., Wang, T., Xuan, Z. (2019). A novel intercooled series expansion refrigeration/liquefaction cycle using pinch technology. Applied Thermal Engineering, 114336. doi:10.1016/j.applthermaleng.2019.114336*

3. Wu, X., Xu, S., & Jiang, M. (2018). *Development of bubble absorption refrigeration technology: A review. Renewable and Sustainable Energy Reviews, 82, 3468–3482.* doi:10.1016/j.rser.2017.10.109

4. Aste, N., Del Pero, C., & Leonforte, F. (2017). *Active refrigeration technologies for food preservation in humanitarian context – A review. Sustainable Energy Technologies and Assessments, 22, 150–160.* doi:10.1016/j.seta.2017.02.014

5. Capra, F., Magli, F., & Gatti, M. (2019). *Biomethane liquefaction: A systematic comparative analysis of refrigeration technologies. Applied Thermal Engineering, 158, 113815.* doi:10.1016/j.applthermaleng.2019.113815

6. Rodríguez, D., Bejarano, G., Vargas, M., Lemos, J. M., & Ortega, M. G. (2019). *Modelling and cooling power control of a TES-backed-up vapour-*

compression refrigeration system. *Applied Thermal Engineering, 114415.* doi:10.1016/j.applthermaleng.2019.114415

7. Bhattad, A., Sarkar, J., & Ghosh, P. (2018). *Improving the performance of refrigeration systems by using nanofluids: A comprehensive review. Renewable and Sustainable Energy Reviews, 82, 3656–3669.* doi:10.1016/j.rser.2017.10.097

8. Bian, Y., Pan, J., Liu, Y., Zhang, F., Yang, Y., & Arima, H. (2019). *Performance analysis of a combined power and refrigeration cycle. Energy Conversion and Management, 185, 259–270.* doi:10.1016/j.enconman.2019.01.072

9. Papadopoulos, A. I., Kyriakides, A.-S., Seferlis, P., & Hassan, I. (2019). Absorption refrigeration processes with organic working fluid mixtures- a review. Renewable and Sustainable Energy Reviews, 109, 239–270. doi:10.1016/j.rser.2019.04.016

10. Vithya, P., Sriram, G., & Arumugam, S. (2019). *Effect of Biodegradable Refrigeration oil on the Tribological Behaviour of Liner/Ring Tribo pair material of Hermetically Sealed Compressors. Materials Today: Proceedings, 16, 488–495.* doi:10.1016/j.matpr.2019.05.120

11. Gullo, P., Tsamos, K., Hafner, A., Ge, Y., & Tassou, S. A. (2017). State -of-the-art technologies for transcritical R744 refrigeration systems – a theoretical assessment of energy advantages for European food retail industry. Energy Procedia, 123, 46–53. doi:10.1016/j.egypro.2017.07.283

12. Gao, Y., & Gao, T. (2019). Simulation study on the performance of direct expansion geothermal refrigeration system using carbon dioxide transcritical cycle. Energy Procedia, 158, 5479–5487. doi:10.1016/j.egypro.2019.01.598

13. Mohammadi, K., Efati Khaledi, M. S., & Powell, K. (2019). *A novel hybrid dual-temperature absorption refrigeration system: Thermodynamic, economic, and environmental analysis. Journal of Cleaner Production.* doi:10.1016/j.jclepro.2019.06.130

14. Liu, Y., & Yu, J. (2019). *Performance evaluation of an ejector subcooling refrigeration cycle with zeotropic mixture R290/R170 for low-temperature freezer applications. Applied Thermal Engineering, 161, 114128.* doi:10.1016/j.applthermaleng.2019.114128

15. Cui, P., Yu, M., Liu, Z., Zhu, Z., & Yang, S. (2019). Energy, exergy, and economic (3E) analyses and multi-objective optimization of a cascade absorption refrigeration system for low-grade waste heat recovery. Energy Conversion and Management, 184, 249–261. doi:10.1016/j.enconman.2019.01.047

16. Rostamnejad, H., & Zare, V. (2019). *Performance improvement of ejector expansion refrigeration cycles employing a booster compressor using different refrigerants: Thermodynamic analysis and optimization. International Journal of Refrigeration.* doi:10.1016/j.ijrefrig.2019.02.031

17. Lu, Y., Roskilly, A. P., & Ma, C. (2017). A techno-economic case study using heat driven absorption refrigeration technology in UK industry. Energy Procedia, 123, 173–179. doi:10.1016/j.egypro.2017.07.254

18. Khanmohammadi, S., Kizilkan, O., & Ahmed, F. W. (2019). *Tri-objective optimization of a hybrid solar-assisted power-refrigeration system working with supercritical carbon dioxide. Renewable Energy.* doi:10.1016/j.renene.2019.11.155

19. Lu, Y., Roskilly, A. P., Huang, R., & Yu, X. (2019). *Study of a novel hybrid refrigeration system for industrial waste heat recovery. Energy Procedia, 158, 2196–2201.* doi:10.1016/j.egypro.2019.01.620

20. Rodríguez-Muñoz, J. L., & Belman-Flores, J. M. (2014). *Review of diffusion–absorption refrigeration technologies. Renewable and Sustainable Energy Reviews, 30, 145–153.* doi:10.1016/j.rser.2013.09.019

21. Besagni, G., Mereu, R., & Inzoli, F. (2016). *Ejector refrigeration: A comprehensive review. Renewable and Sustainable Energy Reviews, 53, 373–407.* doi:10.1016/j.rser.2015.08.059

22. Xu, X., Li, Y., Yang, S., & Chen, G. (2017). *A review of fishing vessel refrigeration systems driven by exhaust heat from engines. Applied Energy, 203, 657–676.* doi:10.1016/j.apenergy.2017.06.019

23. Suamir, I. N., Rasta, I. M., Sudirman, & Tsamos, K. M. (2019). Development of Corn-Oil Ester and Water Mixture Phase Change Materials for Food Refrigeration Applications. Energy Procedia, 161, 198–206. doi:10.1016/j.egypro.2019.02.082

24. Zolfagharkhani, S., Zamen, M., & Shahmardan, M. M. (2018). *Thermodynamic analysis and evaluation of a gas compression refrigeration cycle for fresh water production from atmospheric air. Energy Conversion and Management, 170, 97–107.* doi:10.1016/j.enconman.2018.05.016

25. Keppler, D. (2018). Absorption chillers as a contribution to a climate-friendly refrigeration supply regime: Factors of influence on their further diffusion. Journal of Cleaner Production, 172, 1535–1544. doi:10.1016/j.jclepro.2017.10.276

26. Mylona, Z., Kolokotroni, M., Tsamos, K. M., & Tassou, S. A. (2017). *Comparative analysis on the energy use and environmental impact of different refrigeration systems for frozen food supermarket application. Energy Procedia, 123, 121–130.* doi:10.1016/j.egypro.2017.07.234

27. Fan, Y., Luo, L., & Souyri, B. (2007). *Review of solar sorption refrigeration technologies: Development and applications. Renewable and Sustainable Energy Reviews, 11(8), 1758–1775.* doi:10.1016/j.rser.2006.01.007

28. Wang, D. C., Li, Y. H., Li, D., Xia, Y. Z., & Zhang, J. P. (2010). *A review on adsorption refrigeration technology and adsorption deterioration in physical adsorption systems. Renewable and Sustainable Energy Reviews, 14(1), 344–353.* doi:10.1016/j.rser.2009.08.001

29. Sarbu, I., & Sebarchievici, C. (2015). *General review of solar-powered closed sorption refrigeration systems. Energy Conversion and Management, 105, 403–422.* doi:10.1016/j.enconman.2015.07.084

30. Abdulateef, J. M., Sopian, K., Alghoul, M. A., & Sulaiman, M. Y. (2009). *Review on solar-driven ejector refrigeration technologies. Renewable and Sustainable Energy Reviews, 13(6-7), 1338–1349.* doi:10.1016/j.rser.2008.08.012

31. Bellos, E., & Tzivanidis, C. (2017). *Optimum design of a solar ejector refrigeration system for various operating scenarios. Energy Conversion and Management, 154, 11–24.* doi:10.1016/j.enconman.2017.10.057 ,

32. Franco, V., Blázquez, J. S., Ipus, J. J., Law, J. Y., Moreno-Ramírez, L. M., & Conde, A. (2018). *Magnetocaloric effect: From materials research to refrigeration devices. Progress in Materials Science, 93, 112–232.* doi:10.1016/j.pmatsci.2017.10.005

33. Logesh, K., Baskar, S., Azeemudeen, M., Praveen Reddy, B., & Venkata Subba Sai Jayanth, G. (2019). *Analysis of Cascade Vapour Refrigeration System with Various Refrigerants. Materials Today: Proceedings, 18, 4659–4664.* doi:10.1016/j.matpr.2019.07.450

34. Dhahad, H. A., Hussen, H. M., Nguyen, P. T., Ghaebi, H., & Ashraf, M. A. (2020). *Thermodynamic and thermoeconomic analysis of innovative integration of Kalina and absorption refrigeration cycles for simultaneously cooling and power generation. Energy Conversion and Management, 203, 112241.* doi:10.1016/j.enconman.2019.112241

35. Ghorbani, B., Shirmohammadi, R., Amidpour, M., Inzoli, F., & Rocco, M. (2019). Design and thermoeconomic analysis of a multi-effect desalination unit equipped with a cryogenic refrigeration system. Energy Conversion and Management, 202, 112208. doi:10.1016/j.enconman.2019.112208

36. Ghoreishi-Madiseh, S. A., Kuyuk, A. F., Kalantari, H., & Sasmito, A. P. (2019). Ice versus battery storage; a case for integration of renewable energy in refrigeration systems of remote sites. Energy Procedia, 159, 60–65. doi:10.1016/j.egypro.2018.12.018

37. Patel, B., Kachhwaha, S. S., & Modi, B. (2017). *Thermodynamic Modelling and Parametric Study of a Two Stage Compression-Absorption Refrigeration System for Ice Cream Hardening Plant. Energy Procedia, 109, 190–202.* doi:10.1016/j.egypro.2017.03.091

38. Cerri, D., Luglietti, R., Rosa, P., & Terzi, S. (2016). Lifecycle Optimization in the Refrigeration Industry: A Decision-support Simulation Toolbox (DSST). Procedia CIRP, 48, 277–282. doi:10.1016/j.procir.2016.03.140

39. Rai, A., & Tassou, S. A. (2017). *Energy demand and environmental impacts of alternative food transport refrigeration systems. Energy Procedia, 123, 113–120.* doi:10.1016/j.egypro.2017.07.267

40. Fito, J., Coronas, A., Mauran, S., Mazet, N., & Stitou, D. (2018). *Definition and performance simulations of a novel solar-driven hybrid absorption-thermochemical refrigeration system. Energy Conversion and Management, 175, 298–312.* doi:10.1016/j.enconman.2018.08.098

YOUR KNOWLEDGE HAS VALUE

- We will publish your bachelor's and
  master's thesis, essays and papers

- Your own eBook and book -
  sold worldwide in all relevant shops

- Earn money with each sale

Upload your text at www.GRIN.com
and publish for free